Transportation & Communication Series

The History of
Radio

Joanne Mattern

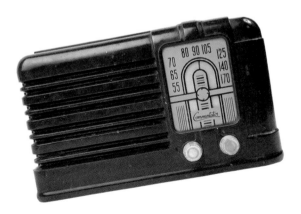

Enslow Publishers, Inc.

40 Industrial Road	PO Box 38
Box 398	Aldershot
Berkeley Heights, NJ 07922	Hants GU12 6BP
USA	UK

http://www.enslow.com

Library of Congress Cataloging-in-Publication Data

Mattern, Joanne, 1963-
    The history of radio / Joanne Mattern.
        p. cm. — (Transportation & communication series)
    Summary: Describes the history of radio, how radios work, careers in
radio, and new developments in the field.
    Includes bibliographical references and index.
    ISBN 0-7660-2027-4
    1. Radio—Juvenile literature. [1. Radio.] I. Title. II. Series.
TK6550.7 .M3797 2002
384.54—dc21
                                    2002009553

Printed in the United States of America

10 9 8 7 6 5 4 3 2 1

To Our Readers: We have done our best to make sure all Internet Addresses in this book were active and
appropriate when we went to press. However, the author and the publisher have no control over and
assume no liability for the material available on those Internet sites or on other Web sites they may link to.
Any comments or suggestions can be sent by e-mail to comments@enslow.com or to the address on the
back cover.

Every effort has been made to locate all copyright holders of material used in this book. If any errors or
omissions have occurred, corrections will be made in future editions of this book.

Illustration Credits: Corel Corporation, pp. 28 (left), 32, 34, 36; Department of Defense,
DVIC, pp. 27, 30; Enslow Publishers, Inc., p. 20; Hemera Technologies, Inc. 1997-2000, pp. 5,
6 (top), 7, 9, 15, 17, 26, 29 (top), 31, 37, 40; Library of Congress, pp. 16, 19, 23; Enslow
Publishers, Inc., pp. 8, 39; Painet Stock Photos, pp. 10, 11, 12, 22, 25, 28 (middle and far
right), 29 (bottom), 33, 35, 38; From the Collections of RMS-Republic.com, pp. 4, 6 (bottom);
© D. Roberts/SPL/Photo Researchers, Inc., p. 13; © SPL/Photo Researchers, Inc., pp. 18, 21,
24.

Cover Illustration: Monster Zero Media using Corel Corporation background images and
© D. Roberts/SPL/Photo Researchers, Inc.

Contents

Chapter 1

"Come Quickly! Danger!"

It was a foggy night in January 1909. The R.M.S. *Republic* was sailing on the North Atlantic Ocean. Suddenly, there was a terrible crash. The *Republic* had hit another ship, the S.S. *Florida*. Both ships were damaged. The *Republic* was in bad shape. Its passengers and crew climbed on board the *Florida*. Then everyone waited for help to arrive. Until that night, waiting for help might have taken days or weeks. Ships had no way to communicate with each other over long distances. Stranded passengers had to hope another ship would come by.

But ships had started to carry a new

From the R.M.S. *Republic*'s radio room (left), Jack Binns called for help. R.M.S. means Royal Mail Ship.

invention called the wireless telegraph, or the radio. The radio let messages be sent to other ships and to stations on land.

Jack Binns was the *Republic*'s wireless telegraph operator. Right after the accident, Binns rushed to the wireless telegraph room. He sent a message that said "CQD." "CQD" stood for "Come Quickly! Danger!"

Binns's message was picked up by the Marconi Siasconsett Station in Massachusetts in the United States. Siasconsett then sent the call for help to all ships in the area. The R.M.S. *Baltic* was about 200 miles away from the *Republic*. The *Baltic* came to rescue the people on the *Republic*.

When the *Baltic* got closer, it could not see the *Republic* or the *Florida* in the thick fog. Jack Binns sent wireless messages to the *Baltic* to tell where the *Republic* was. The *Baltic* was able to find the ships and rescue

The passengers and the crew of the *Republic* climbed aboard lifeboats to safety.

6

over four thousand passengers. Soon after, the *Republic* sank in the Atlantic. Only five lives were lost that night. The *Republic-Florida* collision was the first time that radio had been used to save people after a ship accident. Within a few years, radios would be used on ships all around the world. People also had radios in their homes and workplaces. The invention of the radio changed the way we communicate and how we learn about the world.

Thanks to Jack Binns (right) and the wireless telegraph, many people were saved.

The R.M.S. *Baltic* came to the rescue.

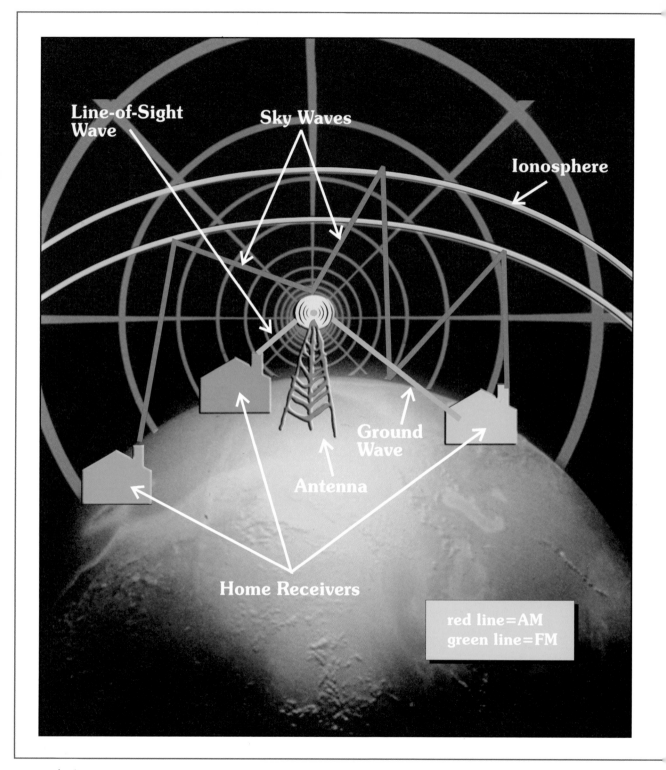

8

Chapter 2

How Radios Work

These are old vacuum tubes.

Radio works by changing sounds into radio waves. Then the waves are changed back into sounds for people to listen to.

Vibrations and Signals

All sounds are made of vibrations. A vibration is a rapid back and forth motion. These vibrations travel through the air in waves called sound waves. Each sound has its own pattern of waves. A low sound, like a truck's motor, makes a longer sound wave. A high sound, like a whistle, makes a smaller sound wave.

To make a radio work, the sound waves must first be changed into electrical signals.

An AM antenna sends out ground waves and sky waves. Ground waves stay close to the ground. Sky waves bounce off the atmosphere and are sent back to earth. An FM antenna sends out line-of-sight waves. That means the antenna can only receive and send out as far as the horizon as seen from the top of the antenna.

 9

This is done by a machine called a transmitter. A transmitter picks up the sound of a person talking or playing music at a radio station. The stronger the transmitter, the farther the radio signals can travel. A radio station with a powerful transmitter can reach more people than a radio station with a weaker transmitter.

The transmitter gathers the sound waves and changes them into electrical signals. Then the transmitter sends, or transmits, the electrical signals to an antenna. Radio waves travel at the speed of light. That is about 186,000 miles per second. Because waves travel so fast, they can be heard at almost the same time they are made, even if they have to travel a long distance.

An antenna is a wire that can send or capture radio signals. Some antennas are placed very high to help them send or get signals as they move through the air. That is why antennas are

Antennas send or capture radio signals.

often seen on top of tall buildings, on tall towers, or even on mountains.

A transmitter antenna broadcasts, or sends, the radio signals out into the air. These signals are picked up by a receiver antenna in radios in people's homes, offices, or cars. The radio receiver then changes the signals back into sounds. When you turn on the radio, you hear music or speech.

What is the Frequency?

Radio programs are broadcast from radio stations. There are thousands of radio stations in the world. Many of them are in the United States. Each radio station broadcasts at a different frequency or channel. Frequency is the number of cycles per second of a radio wave. These frequencies are assigned in the United States by the federal government. Using different frequencies means that radio stations will not interfere with each other's broadcasts. If you tune your radio to a specific

Antennas are usually placed on top of tall buildings, towers, or even mountains.

frequency, such as 97.5 FM, you will hear the radio station that broadcasts on that frequency. To hear a different station, you have to tune in a different frequency.

AM and FM

There are two kinds of radio transmissions, AM and FM. AM stands for amplitude modulation. Modulation means change. AM radio waves change amplitude, or strength, to match changes in the signal coming from the radio station. AM radio broadcasts can travel over long distances. They sometimes are not very clear because of static, or crackling noises from electricity in the air.

Radios can be tuned to FM and AM stations.

FM stands for frequency modulation. FM radio waves are always the same amplitude. The frequency of the waves is changed to send the signal from the radio station. FM broadcasts cannot travel as far as AM broadcasts. But

FM broadcasts do have less static and better sound quality.

The Parts of a Radio

All radios have five basic parts. They are the antenna, the tuner, the intermediate-frequency amplifier and detector, the audio-frequency processor and amplifier, and the speaker.

The antenna is a piece of wire or metal that picks up radio waves. Some radios have antennas built inside the radio. Others, like a car radio, have the antenna sticking out.

The tuner lets the user tune the radio to certain frequencies. To listen to a radio station that broadcasts on 95.5 FM, for example, you would tune your radio to that frequency.

The intermediate-frequency amplifier receives the signal from the tuner. The amplifier makes the signal stronger. Then the audio-frequency processor and amplifier makes the signal even better. You can adjust

Speakers change the electrical signal back into sound. This X ray of a small radio shows the electronic parts inside the radio. The big circle at the top is the speaker.

 13

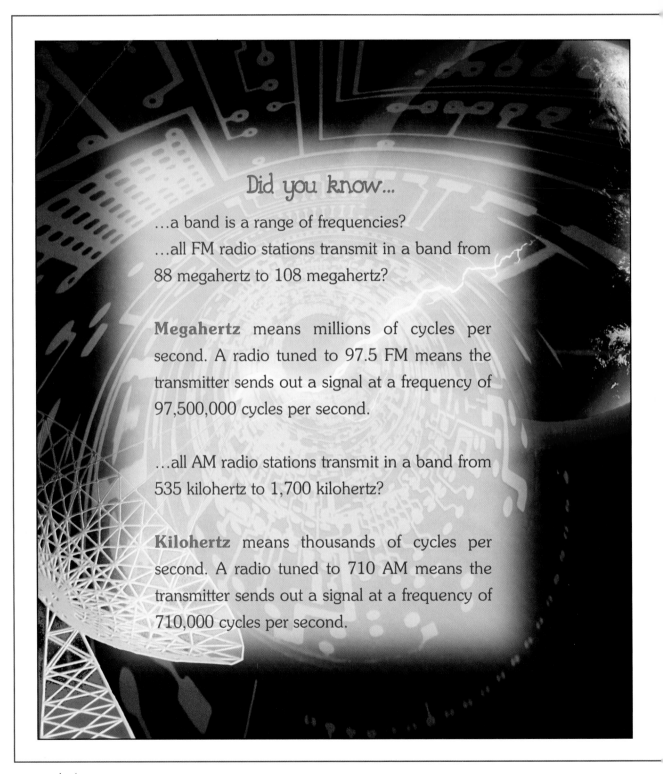

Did you know...

...a band is a range of frequencies?

...all FM radio stations transmit in a band from 88 megahertz to 108 megahertz?

Megahertz means millions of cycles per second. A radio tuned to 97.5 FM means the transmitter sends out a signal at a frequency of 97,500,000 cycles per second.

...all AM radio stations transmit in a band from 535 kilohertz to 1,700 kilohertz?

Kilohertz means thousands of cycles per second. A radio tuned to 710 AM means the transmitter sends out a signal at a frequency of 710,000 cycles per second.

the audio-frequency by turning the volume up or down. You can also change the settings for the bass or treble. Bass is the lower part of a sound. Treble is the higher part of a sound.

Finally, the signal is sent to the radio's speakers. The speakers change the electrical signal back into sound. Electrical signals pass through a voice coil connected to a magnet. The signals move through the coil and make the magnet vibrate. These vibrations create sound waves. These sound waves come out of the speaker to your ear. The radio broadcast can be heard at almost the same time it is being sent from a radio station many miles away.

Speakers allow you to hear the radio broadcast.

Chapter 3

The History of Radio

Radio has been part of our world for over one hundred years. The beginnings of radio go back to the 1800s.

Discovering Radio Waves

During the 1830s, two scientists were working separately with magnets. They were an American scientist named Joseph Henry and a British scientist named Michael Faraday. Each scientist discovered that sending a current into one wire could make a current in another wire. This would happen even if the two wires were not connected. They did not know why this happened.

For over one hundred years, radio has been entertaining and informing people. This old radio (left) is from the 1920s.

Heinrich Hertz discovered electromagnetic waves.

In 1864, a British scientist named James Clerk Maxwell thought of an answer. Maxwell said there were radio waves that traveled in straight lines at the speed of light. Although nobody could see the waves, they could send an electrical current from one wire to another.

Around 1888, Maxwell's theory was proved by a German physicist, Heinrich Hertz. Frequencies are now measured in units of what are called hertz.

Marconi's Marvel

A young Italian man named Guglielmo Marconi made wireless technology even better. Marconi read about what Hertz had done. In 1895, Marconi found a way to use radio waves to send sounds. He built a simple transmitter and receiver out of wires and metal. He used the transmitter and receiver to send a signal about one mile. He sent the message by using Morse code.

Marconi called his radio the wireless

telegraph. He kept trying to make it better. He was able to send the signals farther. In 1901, he sent signals 1,700 miles across the Atlantic Ocean. People were excited about what Marconi had done and what it would mean to communication. Marconi's invention soon spread around the world. It was first used on ships. Wireless equipment was placed on ships and in stations on land. This let ships communicate with each other and with people on shore using Morse code.

Marconi invented special transmitters and receivers that were able to capture and send radio waves. He called his invention the wireless telegraph.

Guglielmo Marconi (1874–1937)

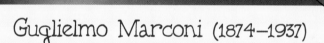

Guglielmo Marconi was born in Bologna, Italy, in 1874. He came from a rich family and had a good education. In 1894, Marconi was studying to be an electrical engineer. He began experimenting with wireless. He invented a special transmitter and receiver to capture and send radio waves. He also used an amplifier to make the signals stronger. Marconi spent the rest of his life making wireless telegraphy better. In 1909, he won the Nobel Prize in physics. Marconi shared his prize with a German named Karl Ferdinand Braun. Braun had changed Marconi's transmitter so it could send signals even farther.

Marconi, on the far right, spent his life making wireless telegraphy better.

Wireless telegraphy let people communicate using Morse code. Morse code uses dits (dots or ●) and dahs (dashes or –) that stand for letters and numbers. To send a message, an operator like Jack Binns would put together the dits and dahs.

Morse code uses short and long sounds combined in various ways to stand for letters, numerals, and other characters. A short sound is called a dit. A long sound is called a dah. People who understood Morse code had to decode the dits and dahs and turn them back into words. It was like working a puzzle.

Being able to send a message over the radio saved many lives.

a= ● –	m= – –	y= – ● – –
b= – ● ● ●	n= – ●	z= – – ● ●
c= – ● – ●	o= – – –	1= ● – – – –
d= – ● ●	p= ● – – ●	2= ● ● – – –
e= ●	q= – – ● –	3= ● ● ● – –
f= ● ● – ●	r= ● – ●	4= ● ● ● ● –
g= – – ●	s= ● ● ●	5= ● ● ● ● ●
h= ● ● ● ●	t= –	6= – ● ● ● ●
i= ● ●	u= ● ● –	7= – – ● ● ●
j= ● – – –	v= ● ● ● –	8= – – – ● ●
k= – ● –	w= ● – –	9= – – – – ●
l= ● – ● ●	x= – ● ● –	0= – – – – –

A Growing Industry

At first, radio signals were not very strong. Then, in 1904, British scientist Sir John Fleming built the first diode. A diode uses electricity to find and send radio signals. In 1906, Lee De Forest built what he called the Audion to make radio signals stronger. The Audion was a type of vacuum tube. These tubes took radio signals and turned the electrons into stronger waves. De Forest's Audion became an important part of radio. It was also important in radar, television, and computers. During the 1950s, Audions were replaced by the much smaller transistor.

Lee De Forest is standing next to one of his many inventions, a movie camera.

Lee De Forest (1873–1961)

Lee De Forest was born in Council Bluffs, Iowa, in 1873. He went to Yale University. De Forest designed many different transmitters. The Audion was his most important invention. De Forest also worked on introducing sound to motion pictures, or movies. He invented more than 300 electronic devices for radio and movies.

Golden Age of Broadcasting

The years between 1920 and 1950 were called the Golden Age of Broadcasting. By the 1920s, wireless had become radio, with stations reaching many listeners. Radios were being used for many more things. In 1920, KDKA in Pittsburgh, Pennsylvania, became one of the first radio stations to have regular broadcasts. About 1,000 people in the area listened to the broadcasts every week.

As more radio stations went on the air,

The 1920s were the start of the Golden Age of Broadcasting. Many people enjoyed listening to the radio.

people began to listen to radio for news and fun. For the first time, people could hear baseball games. People listened to popular comedians such as Abbott and Costello. And they listened to mysteries and dramas such as *The Green Hornet* and *Superman*. During the day, people listened to soap operas such as *The Guiding Light*. They were called soap operas because many soap companies paid for the programs.

Popular bands and musicians such as Glenn Miller, Benny Goodman, and Duke Ellington broadcast concerts over the radio. Families sat around the radio every night to listen to these programs.

Radio also brought important news events right into people's houses. On May 6, 1937, a German airship called the *Hindenburg* crashed and burned while landing in Lakehurst, New Jersey. The next day listeners heard a recording of radio reporter Herb Morrison tell about the disaster.

Duke Ellington, seen here, was one of the first musicians to be heard on radio.

Edwin Armstrong introduced FM radio.

Another exciting radio moment occurred on October 30, 1938. Orson Welles broadcast a radio show called *War of the Worlds*. *War of the Worlds* was based on a book with the same title by H. G. Wells. This story said that Martians had landed and were taking over the country. Many listeners thought the story was true. People panicked all over the United States. The next day, Welles apologized for scaring everybody.

Radio was very important during World War II. Americans sat around the radio to hear news about the battles. They also heard President Franklin D. Roosevelt talk to the

Edwin Armstrong (1890–1954)

Edwin Howard Armstrong was born in New York City in 1890. He was an American electrical engineer. He played an important part in the development of radio. In 1933, Armstrong introduced FM radio. FM provided better sound quality and less static than AM. He also developed several radio receivers.

nation during this time. These talks were called "fireside chats."

Radios Get Smaller...and Better!

Early radios had become large pieces of furniture. These radios had to be large and bulky. They needed to hold all the parts needed to receive signals.

In 1947, scientists at the Bell Telephone Laboratories invented the transistor to amplify radio signals. Transistors are solid and much smaller than vacuum tubes, which looked like light bulbs. Radios became smaller. During the 1950s and 1960s, many people carried transistor radios in their pockets. These radios were powered by batteries, so people could listen to them wherever they went.

During the 1970s, a small radio with earphones became popular. This radio was

During World War II, radio was important. Americans were able to hear news about the battles. Soldiers were able to use radio, like the one seen here, to communicate.

Radios slowly became smaller during the late 1940s.

Small radios with headphones became very popular during the 1970s.

called the Sony Walkman™. You could listen to it while walking around.

Better equipment continued to make radio signals stronger. Sound quality also got better. In 1961, FM radio began broadcasting in stereo. Stereo AM broadcasts began in 1982. Stereo broadcasts use two different channels to make more natural sounds.

In the late 1980s, digital audio broadcasting was developed. Digital broadcasting made sound much sharper and clearer. It is like comparing the sound quality of a CD to a cassette tape.

Point-to-Point Communication

Radio is not only used to broadcast music, news, and sports. People also use radio to make their jobs easier and to help others. This is called point-to-point communication.

Police officers and firefighters carry two-way radios. Portable two-way radios are called walkie-talkies. Other people who help in emergencies carry radios, too. These radios let workers talk to a command center and to each other. The command center sends instructions on how to help in an emergency. Taxi and bus drivers also use two-way radios. They can get instructions on where to pick up passengers. Construction workers use two-way radios to talk to other workers. Soldiers use them to talk to their base.

Soldiers use two-way radios to communicate with their bases and with each other. Here, a communications station at a field hospital is being set up during Operation Desert Storm.

Firefighters, security officers, police, and forest rangers use two-way radios to keep in touch.

Another popular radio is the ham radio. Ham radios are small radio transmitters and receivers. People use their radios to talk to others all over the world. Many people keep a list of how many different people in other countries they have talked to.

People use them just for fun. But, ham radio operators are also important during emergencies. If there has been an earthquake, tornado, or other disaster, telephone lines might have been destroyed. Ham radio operators are able to send messages to and from the site. They can tell family members if their loved ones are safe. They can also get help for people who need it.

Many people thought that television would be the end of radio. That has not happened. Today, millions of people listen to music, sports, talk shows, and news on the radio. Radio still connects people every day.

This is a ham radio.

Many people today enjoy listening to radio for news, music, and sports.

Chapter 4

Careers in Radio

It takes many people to put together a radio broadcast. The invention of radio made a whole new industry and many new jobs.

On the Air

Many people work in broadcasting. Disc jockeys (DJs) play music. Talk show hosts interview people during the broadcast, which is also called "being on the air." News announcers report on important events. Weather forecasters report on weather conditions. Sportscasters tell listeners about sporting events. Traffic reporters tell listeners about any problems on the roads. They also tell

The invention of radio created a whole new set of jobs. This woman is a DJ.

Even though television lets people see events as they happen, radio lets people hear about them.

if buses and trains are running on time.

All these people work in a part of the radio station called the studio. The studio has a lot of electronic equipment. Announcers or DJs speak into a microphone. The microphone carries the sound of their voices to the transmitter. Usually, a sound engineer is also in the studio. The engineer makes sure all the equipment is working and that the signals are broadcast in the right way.

Some radio station studios look like this.

Behind the Scenes

Many more people work behind the scenes in radio. Reporters gather information about news stories. Writers prepare scripts for announcers to read on the air. Program directors decide what type of material to include in the broadcasts.

This woman is ready to go on the air.

Other people take care of the technical parts of radio broadcasting. Technicians and engineers operate the equipment and computers used to broadcast a program. Maintenance people fix equipment and make sure everything is working the way it should.

Sales people sell air time to companies who want to advertise their businesses. The

Sales people sell air time to companies for their advertisements.

marketing department looks for ways to get more people to listen to the radio station. These people set up special events for listeners, such as concerts.

At larger radio stations, these jobs are divided up among many different people. At smaller stations, one person may have many different jobs. For example, the same person might report a story, write the script, and read the story on the air.

Radio stations have a lot of electronic equipment in them that help the announcers do their job.

These teens can listen to some radio stations using their computers and the Internet.

Chapter 5

The Future of Radio

Today, we can watch television, listen to CDs, or use computers to surf the Internet. But, radio is still an important part of our lives. Most homes in the United States have several radios. Almost every car has a radio, too. People turn on the radio every day to hear news, music, or to find out how their favorite sports teams are doing.

New Technology

One reason radio is still popular is because technology is improving. Radio's sound quality and signal strength are getting better. Radio is also being broadcast in different ways.

Satellite technology (left) helps to improve the quality of radio signals.

With all the new technology, radio is still an important part of people's lives.

Today, you can listen to the radio over the Internet. Broadcasting over the Internet allows a station to reach listeners all over the world.

Satellite Radio

A new technology is satellite radio. Satellite radio was first put in cars in September 2001. These cars have a special receiver. Its antenna captures signals sent from a satellite orbiting high above the earth. Satellites let a broadcast be heard anywhere in the country. A person driving from New York to California could listen to the same station during the entire trip. Listeners can also choose from a wider selection of music. One hundred years ago,

Some cars have satellite radio. A person can drive from New York City to Los Angeles, California, while listening to the same radio station.

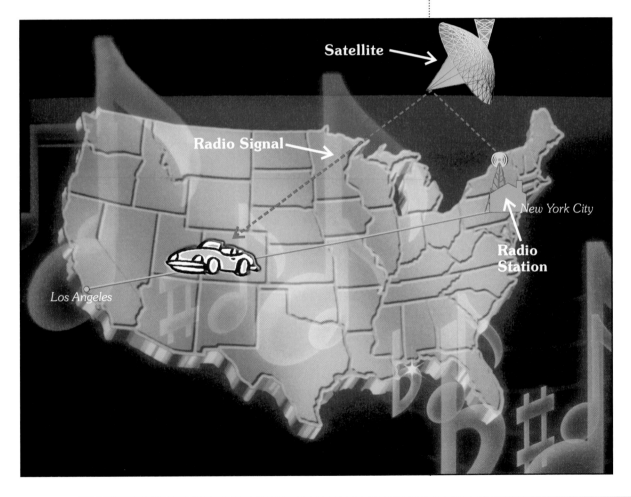

people read newspapers or listened to local gossip to find out what was going on. It took news days or even weeks to travel from one place to another. Today, you can just turn on the radio to hear the latest news, sports, weather, and music. The radio has helped everyone get in touch with the world.

Radio is still an important part of our lives.

Timeline

1830s—American scientist Joseph Henry and British scientist Michael Faraday discover that sending a current into one wire can make a current in another wire, even if the two wires are not connected.

1864—British scientist James Clerk Maxwell says that radio waves travel at the speed of light.

1888—Heinrich Hertz proves electromagnetic waves exist and act like light.

1895—Italian inventor Guglielmo Marconi sends radio signals a few miles through the air.

1901—Marconi sends radio signals 1,700 miles across the Atlantic Ocean.

1904—British scientist Sir John Fleming invents the diode.

Timeline

1906—American scientist Lee De Forest invents the Audion.

1920—KDKA in Pittsburgh becomes one of the first radio stations to make regular broadcasts.

1920s–1950s—Radio becomes a form of entertainment.

1947—Scientists at Bell Telephone Laboratories invent the transistor.

1961—FM radio begins broadcasting in stereo.

1982—AM radio begins broadcasting in stereo.

2001—Satellite radio is introduced.

Words to Know

amplifier—Something that makes a signal stronger or louder.

amplitude—Strength; Also alternating current or wave.

antenna—A metal device that can receive and send radio signals.

Audion—A device that makes radio signals stronger; now called the triode.

broadcast—A program that is presented on radio or television. Also, to send out by radio.

current—The movement of electricity through a wire.

cycle—A series of events that are repeated over and over.

diode—A device that uses electricity to find and send radio signals.

hertz—The number of cycles produced by a radio wave in one second.

Words to Know

modulation—Change; Also a varying of amplitude for the transmission of information by radio.

receiver—Equipment that gathers radio waves and changes them into sounds.

signal—Electrical pulses that are sent for radio communication.

static—Crackling noises from electricity in the air.

stereo—Using two or more channels so listeners can hear sounds in a more natural way.

Learn More About
Radio

Books

Bridgman, Roger. *Electronics*. New York: Dorling Kindersley Publishing, 2000.

Goldsmith, Mike. *Guglielmo Marconi*. Austin, Tex.: Raintree Steck-Vaughn Publishers, 2002.

Oxlade, Chris. *Radio*. Chicago, Ill.: The Heinemann Library, 2001.

Snow, Panky. *Radio Announcers*. Minnetonka, Minn.: Capstone Press, Inc., 2001.

Learn More About
Radio

Internet Addresses

How Stuff Works: Radio

<http://www.howstuffworks.com/radio.html>

Learn more about how radio works, from simple transmitters to antennas.

History of Radio

<http://history.acusd.edu/gen/recording/radio.html>

This site has a lot of information about the history of radio.

Index

Index